On the Farm

by Annabelle Garges

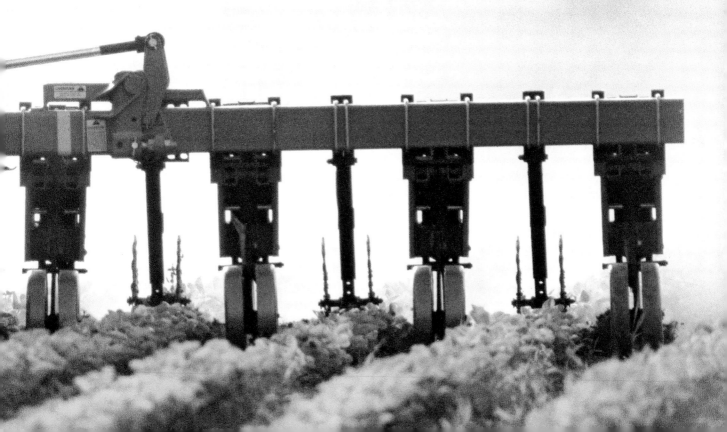

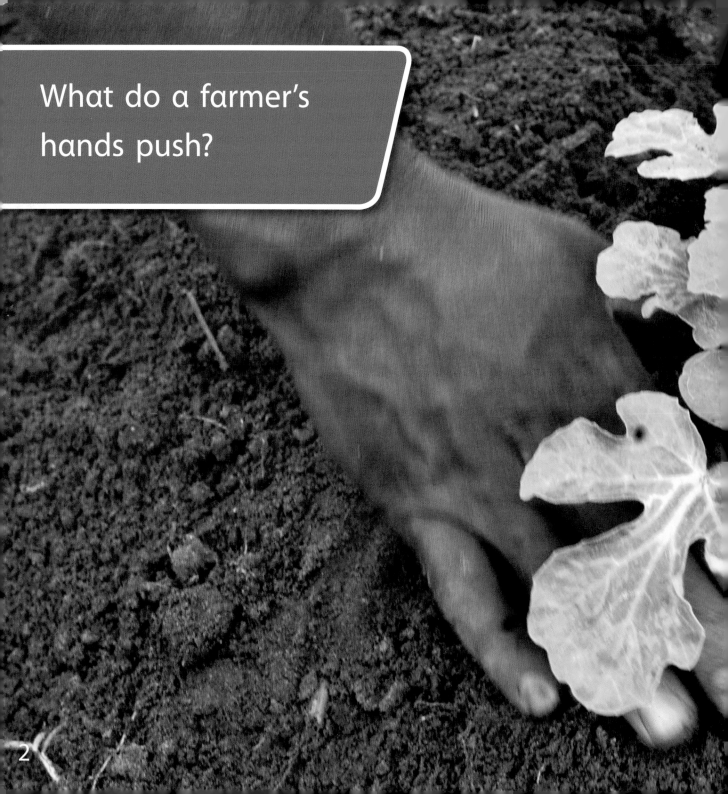

What do a farmer's hands push?

2

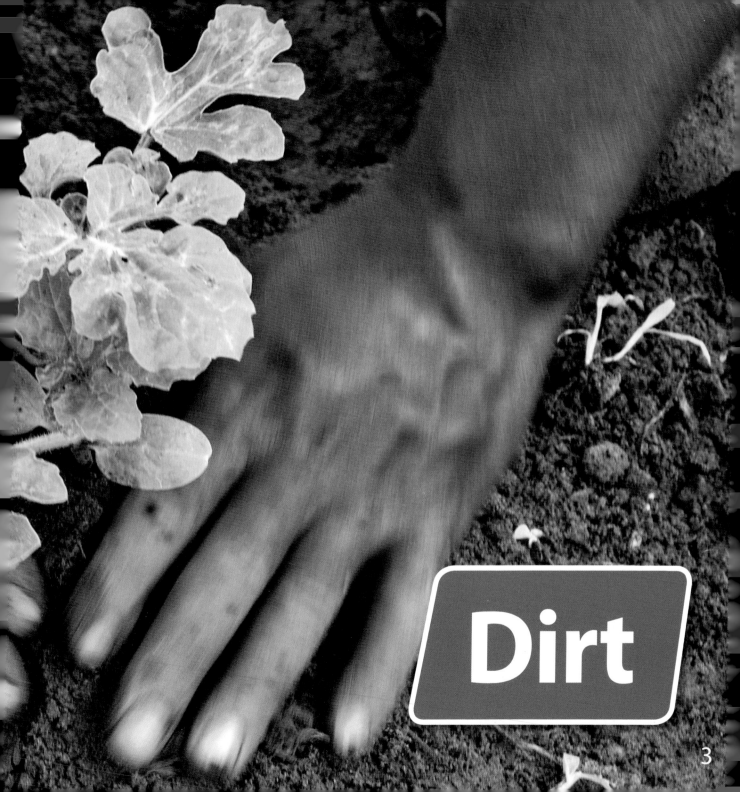

Dirt

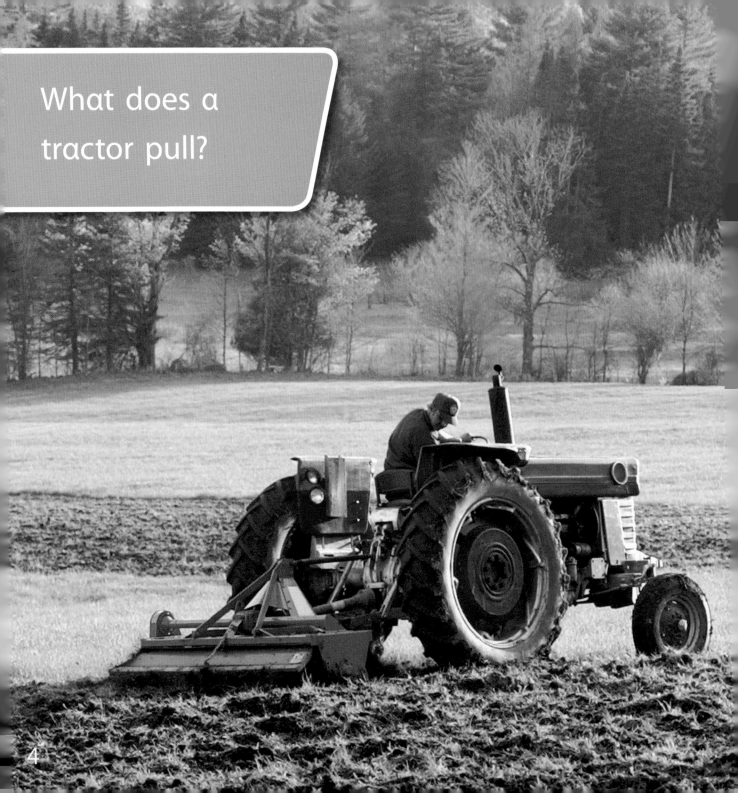

What does a tractor pull?

A plow

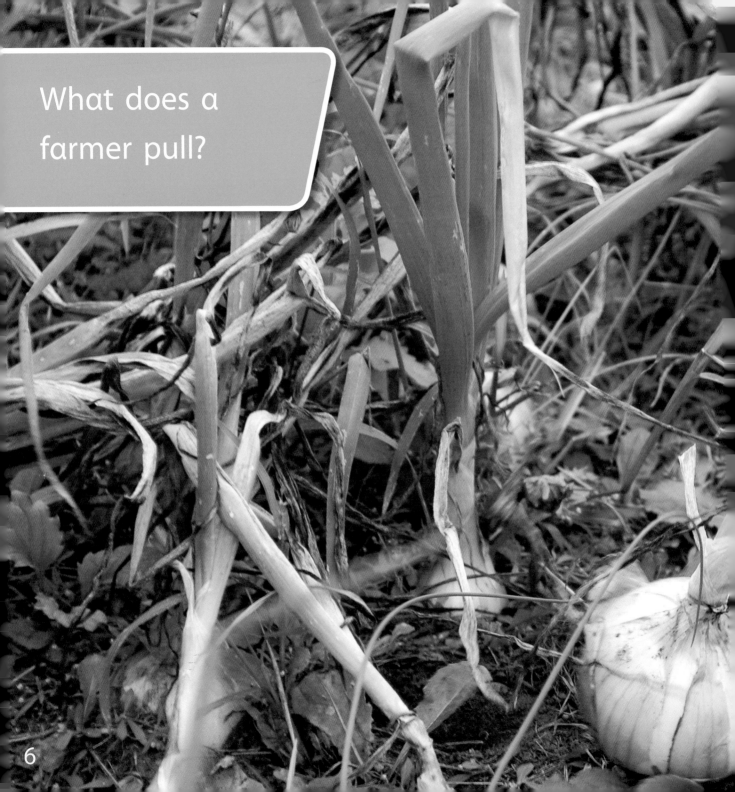

What does a
farmer pull?

6

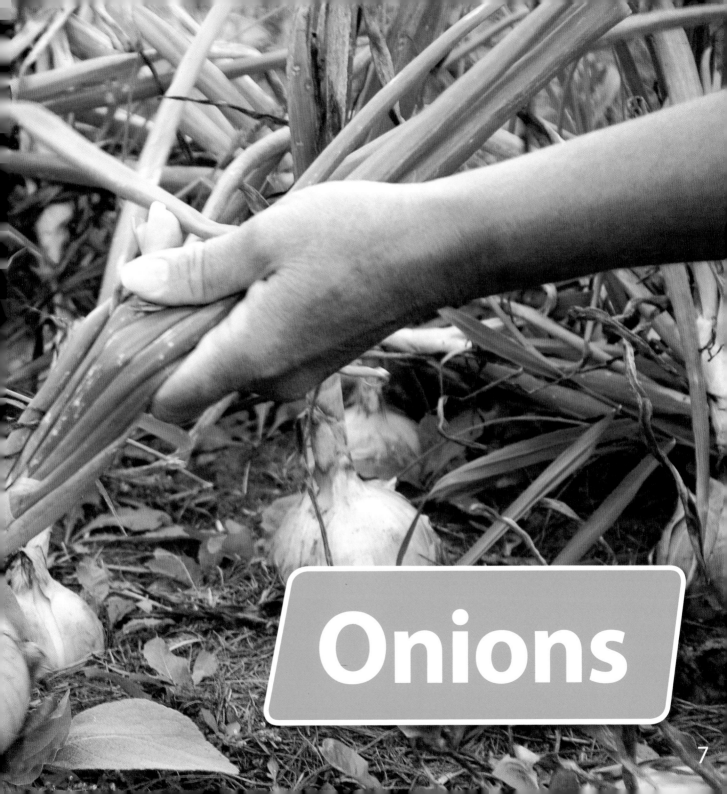

Onions

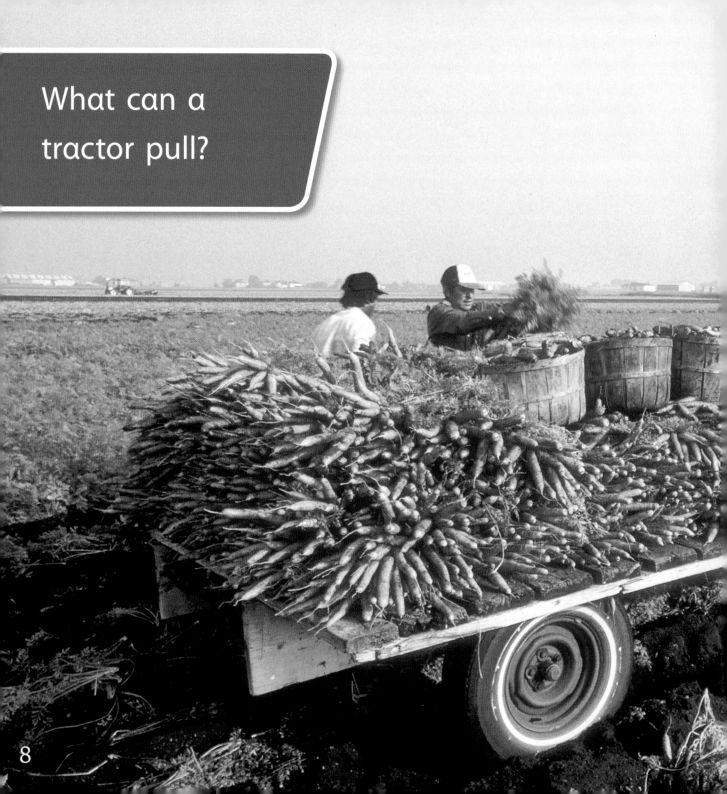

What can a
tractor pull?

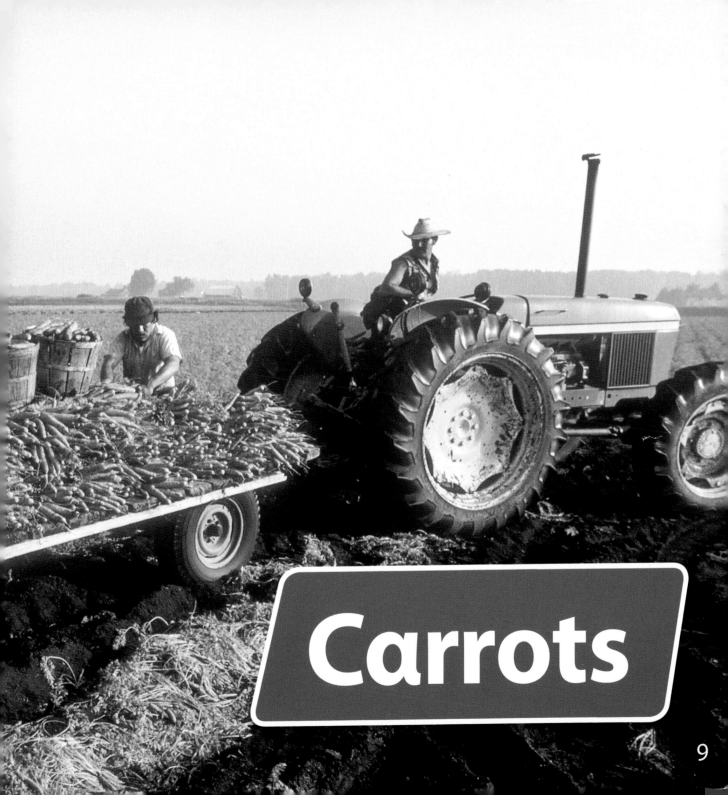

Carrots

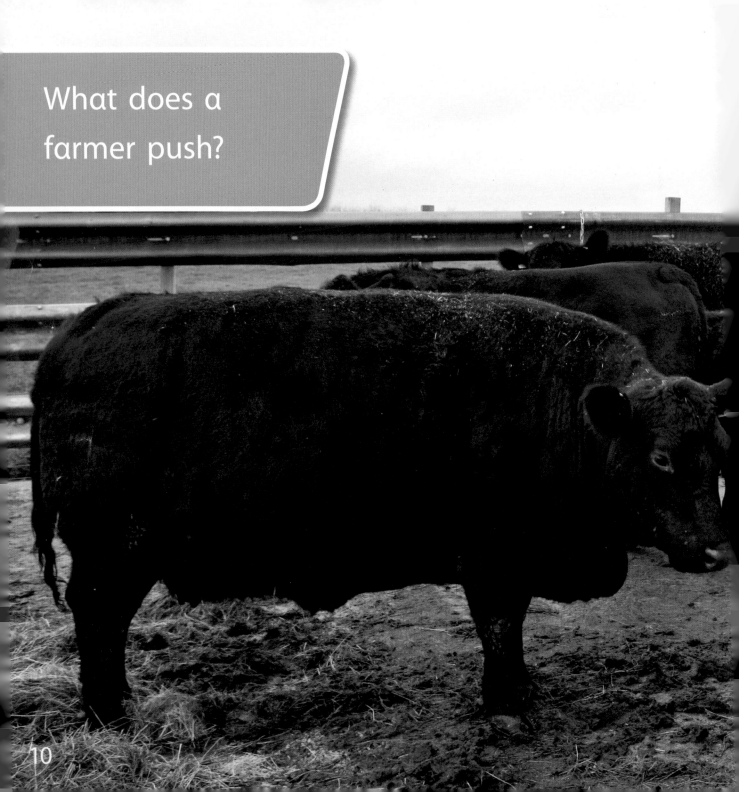

What does a
farmer push?

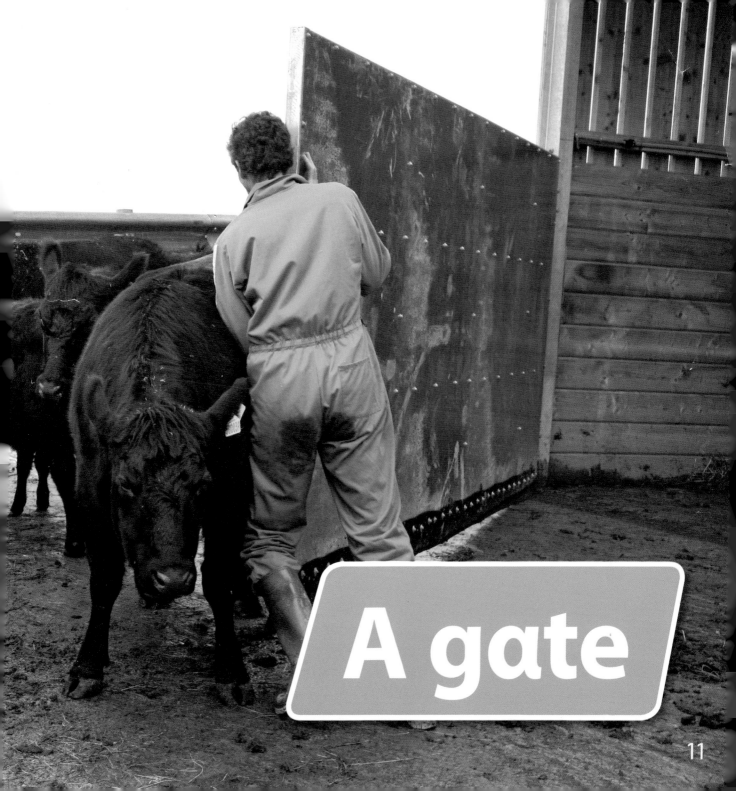

A gate

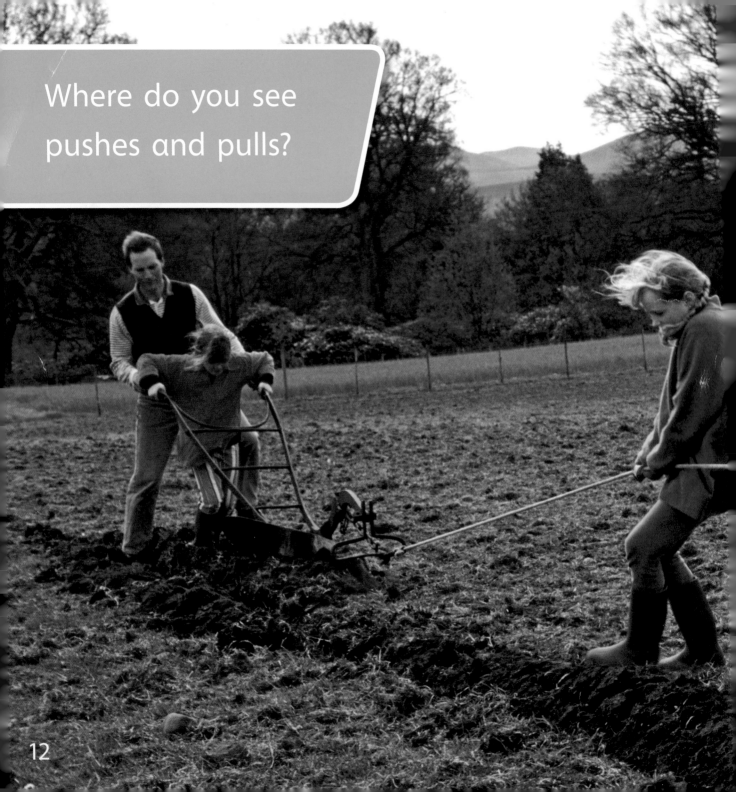

Where do you see pushes and pulls?